P9-AOX-817

DISCARDED
EASTWOODS SCHOOL LIBRARY

SOYBEANS: THE WONDER BEANS

SOYBEANS
THE WONDER BEANS

**by Leonard S. Kenworthy
and Laurence Jaeger**

Illustrated with photographs

JULIAN MESSNER
New York

Published by Julian Messner, a Division of Simon & Schuster, Inc.
1 West 39 Street, New York, N.Y. 10018. All rights reserved.

Copyright © 1976 by Leonard S. Kenworthy and Laurence Jaeger

Design by Ruth Bornschlegel

Library of Congress Cataloging in Publication Data

Kenworthy, Leonard Stout,
    Soybeans: the wonder beans.

    SUMMARY: Introduces the soybean and its many
uses.
    1. Soybean—Juvenile literature. [1. Soybean]
I. Jaeger, Laurence, joint author. II. Title.
SB205.S7K34   635'.655   76-986
ISBN 0-671-32774-7
ISBN 0-671-32775-5 lib. bdg.

150067

• ACKNOWLEDGMENTS

Many persons have assisted the authors in the preparation of this book. Among them were representatives of the United States Department of Agriculture and its extension offices in various states, officials of the Food and Agriculture Organization of the United Nations, and reference librarians. To them and to all the other persons who have helped, the authors extend their thanks.

Three persons have been especially helpful. One was Carol Koch, the editor of the *Soybean Digest,* who supplied the writers with many important leads to sources of information. Another was Wendell Gilbert, a Hoosier soybean farmer, who read and commented upon the manuscript. A third was Dr. H. A. Cate, Associate Professor of Agricultural Extension at the University of Illinois, whose extensive background on soybeans was of great value to the authors of this volume.

Photo Credits

J. C. Allen & Son: pp. 2, 19, 36, 39 (top and bottom), 41, 43, 44 (top and bottom), 47, 48, 49

FAO Photo: p. 55

Food Protein Council: p. 54

William Jaber: p. 52 (top)

National Soybean Processors Association: pp. 30, 37

New York Public Library Picture Collection: p. 27

Nodine Studio: pp. 29, 31, 42

Reprinted by permission from *Newstime,* © 1973 by Scholastic Magazines, Inc.: p. 28

Tennessee Valley Authority: p. 10

U.S. Department of Agriculture: pp. 16, 17, 18, 21, 24

University of Iowa: pp. 34, 46

• CONTENTS

·1·

Introducing
The Wonder Beans

THE STORY OF JACK and the beanstalk doesn't tell what kind of beans Jack dropped on the earth outside his window. But, considering the magic in them, they could have been soybeans.

Of course, soybeans are not magic. But they are special.

There are 2,500 different kinds of soybeans. Most of them are smaller than a pea, though not quite so round. They have tough skins, often with a black or dark brown spot. They come in different colors: yellow, brown, off-white, sometimes nearly black. Some soybeans are a mixture of all these colors. However, it's not their looks that make them special. It's what they do.

Take your breakfast this morning. Did you eat Frosted Flakes, or Cheerios, or some other brand of dry cereal? Soybeans are used to make many breakfast cereals you see lined up on the supermarket shelves.

Soybeans go into the manufacture of all these foods.

Do you love Howard Johnson's ice cream? Or Oreo cookies? Or Aunt Jemima pancakes? Do you stir chocolate or strawberry Quik into your milk? Does your mother bake with Betty Crocker cake mixes?

Every one of those brands has soybeans in them. Soybeans are used to make all brands of ice cream, milk flavorings, mixes, and cookies.

—including these ice cream products!

Does your family use margarine? Do your father and mother drink instant coffee? Do you squirt whipped topping on your dessert? Or pour bottled dressing over your salad? Soybeans are used in making all of these foods, too.

But soybeans are in many other things besides food.

Take this book as an example. Soybeans were used to make the ink that printed the words and pictures.

Soybeans go into the soaps and detergents we use to wash ourselves, our clothes, and our dishes.

Soybeans go into the paints that make our

Soybeans go into paints and glues and jewelry and soap and candles.

homes beautiful, and the finger paints with which you make pictures.

Soybeans go into the fabrics that cover our car seats, the couches in our living rooms, the mattresses on our beds.

Soybeans go into the rubberized boots, coats, and umbrellas that keep us dry when it rains.

Soybeans go into women's makeup and baby lotions.

Soybeans go into the floor tiles we walk on.

Soybeans go into the oils that are used to keep things running smoothly, from bikes to giant machines in factories.

Soybeans go into the feed mixes eaten by chickens, turkeys, cattle, and hogs.

Soybeans even go back into the earth where they grew, to make the soil extra rich and fertile for growing other crops.

These are a lot of different uses for a bean that is a close relative of the common beans we eat: lima beans, kidney beans, navy beans, lentils, peas, and others.

But, then, even Jack was surprised to discover how special his beans really were!

·2·

The Soybean Magic Spreads

THE FIRST AMERICAN to see a soybean was probably a sailor whose ship was visiting China around 1800.

In the United States, soybeans were unknown. But in China they had been raised for thousands of years. Among the Chinese, soybeans were—and still are—an important food.

Perhaps the American sailor was the son of a farm family. What fun, he may have thought, to take home a bag of these strange, dried beans and plant them! That is what he had seen the Chinese farmers do.

But when he planted some soybeans, and other Americans tried it too, they didn't think of them

Sweeping up soybeans after threshing (separating the beans from the pods).

EARLY PHOTOGRAPHS OF SOYBEAN FARMING IN CHINA.

Carts carrying soybeans being loaded on to flatboats on their way to market.

Wheel-like soybean cakes being carried from a warehouse in China.

as food for humans. Maybe it was because of the soybean's tough skin. Or maybe it was because Americans already had other kinds of tasty beans to eat.

Whatever the reason, most American farmers did not grow soybeans at first. Those few who did, raised them as food for their cows, horses, sheep, and hogs.

In the summer, the farmers would let the animals eat the leaves and pods from the growing soybean plants. In the autumn, they would cut off the tops of the plants. These were dried and fed to the animals in the winter when there were no fresh, green things to feed on.

Soon, farmers found their soil wore out less rapidly where soybeans had been planted. They

The roots of a soybean plant. The little balls, or nodules, contain stored nitrogen.

realized the soybeans were keeping the soil more fertile. Today, we understand why soybeans enrich the soil. There's no magic about it.

One of the ways plants grow is by taking food from the soil through their roots. And one of the foods they take from the soil is a gas called *nitrogen*. If nitrogen is not put back into the earth, the soil wears out in a few years.

But soybeans do not take nitrogen from the soil. With help from special bacteria on their roots, they take nitrogen from the air instead. Soybeans are one of the few plants that can do this. That is

Farming in the early 1900s was a lot different from today's methods. Compare this photograph with the one on page 41.

why more and more farmers began to plant them.

One year, farmers would plant a field with soybeans. The next year, they planted corn or cotton or whatever crop they grew for a living. The third year, they went back to soybeans. The fourth year, it was their regular crop again. And so on. This is called *crop rotation*.

By 1900, many farmers had learned that soybeans were good not only as animal feed but also as fertilizer, or as a green-manure crop. Green manure is a crop that is plowed under to bring worn-out soil back to health.

Then developments took place that made soybeans even more important in the United States.

·3·

From Feed for Animals to Food for People— and More!

By 1900, THE UNITED STATES was growing rapidly in population. There were more than 100 million Americans. And every month, thousands of people from other countries were coming here to live.

At the same time, the American way of life was changing. More and more people were living in cities. They needed to be supplied with food. But every year, there were fewer people on the farms to grow food. They, too, were moving to the cities to find jobs in stores, offices, and factories.

So scientists began to think about the soybean

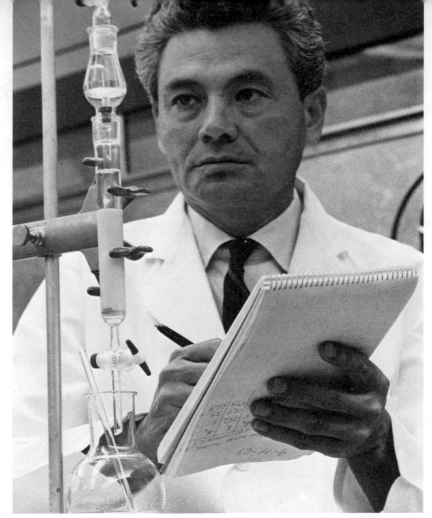

One of the thousands of scientists doing soybean experiments.

as a food for humans. It was already widely grown in the United States. Machines could easily plant and harvest the crop.

If the Chinese had been eating soybeans for

thousands of years, why couldn't Americans be taught to eat them? And if soybeans were fed to cattle and hogs to make them big and healthy, why wouldn't soybeans be good for America's people, too?

Answers to such questions were not long in coming.

By about 1920, mills were at work processing soybeans. Millions of soybeans were changed into oil, soy meal (a powder), and soy flakes.

These soybean products were bought by food companies. They worked out ways to use them in making things that Americans like to eat, such as cereals, candies, breads, cakes, cookies, and crackers.

They also made foods that used soybean products in other ways. One of these new foods was margarine, a substitute for butter. It was made from soybean oil, and it cost a lot less than butter.

Cooking oil and salad oil made of soybeans were other new products of the 1920s. Before that time, they had been made mainly from olives.

Soybean oil was—and still is—much cheaper than olive oil.

By 1930, large numbers of Americans were eating soybean products. Usually these products were mixed into the foods they were already used to eating. Still, it was a big change in the diet of a nation that had always thought of soybeans as food for animals only!

Scientists began looking for even more ways to use soybeans—ways that might not have anything to do with food, or green-manure crops, or animal feed. One of these scientists was George Washington Carver, who was already famous all over the world for his work with the peanut.

In his laboratory in Tuskegee, Alabama, Dr. Carver had made 300 different products from peanuts. A few of the more important are used in the manufacture of ink, soap, flour, paint, dyes, and building materials.

Now, thanks to Dr. Carver and other scientists, soybeans have been found to have as many uses as peanuts. And many large industries depend on

A painting of Dr. George Washington Carver in his laboratory.

soybeans—the food, cosmetic, paint, printing, and paper industries are some.

The soybean industry itself is booming as it provides the oil, flakes, and meal these other in-

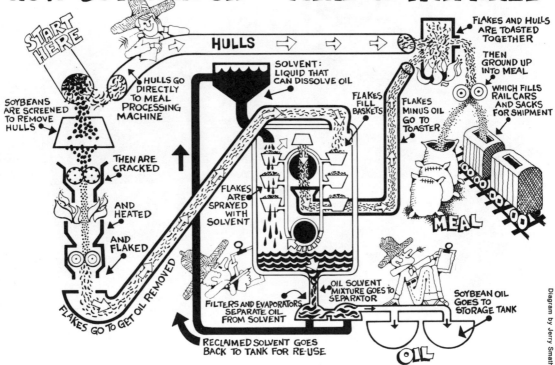

HOW SOYBEAN OIL AND MEAL ARE PREPARED

START HERE

HULLS

FLAKES AND HULLS ARE TOASTED TOGETHER

THEN GROUND UP INTO MEAL

WHICH FILLS RAIL CARS AND SACKS FOR SHIPMENT

HULLS GO DIRECTLY TO MEAL PROCESSING MACHINE

SOLVENT: LIQUID THAT CAN DISSOLVE OIL

FLAKES FILL BASKETS

FLAKES MINUS OIL GO TO TOASTER

SOYBEANS ARE SCREENED TO REMOVE HULLS

THEN ARE CRACKED

AND HEATED

AND FLAKED

FLAKES ARE SPRAYED WITH SOLVENT

MEAL

FLAKES GO TO GET OIL REMOVED

FILTERS AND EVAPORATORS SEPARATE OIL FROM SOLVENT

OIL SOLVENT MIXTURE GOES TO SEPARATOR

SOYBEAN OIL GOES TO STORAGE TANK

RECLAIMED SOLVENT GOES BACK TO TANK FOR RE-USE

OIL

Diagram by Jerry Smath

dustries need. Today, about 130 mills in the United States process soybeans. New machinery is being developed to do the processing faster and better. And scientists are constantly working on ways to

An aerial view of a soybean processing plant.

Inside the plant, huge machines extract soybean oil.

Checking on the machines that process soybeans.

make soybean plants more productive.

A lot of soybeans are needed—where do they all come from?

•4•

How Soybeans Grow

TODAY, THE UNITED STATES grows more soybeans than any other nation. Every year, we grow more than half of the entire world crop. (China is second.)

Soybean farming is carried on in 30 of our 50 states. But no matter where American soybean farmers live, they do their work in much the same way.

In most parts of the United States, the weather is usually right for planting soybeans sometime in May. Southern farmers may be able to plant as early as April. Farmers in the north may have to wait until June.

However, before the seeds can be planted, the farmer must get the soil ready. He attaches a plow to his tractor, and drives through the fields, turning the soil in long straight furrows.

The plow has blades, or shares, which dig into the earth. As the tractor moves along, the shares scoop up the soil and turn it over as it drops

Plowing the soil in preparation for planting.

down again. Each share on the plow turns over long ribbons of soil about 16 inches wide and six inches deep. Sometimes they are wider; sometimes narrower. Some plows can dig up as much as six or eight furrows at a time.

If the farm is large, the plowing may take several days.

When the plowing is done, the farmer goes over the field again, this time with a disk attached to his tractor. As he drives across the field, the blades of the disk chop up the chunks of soil which were left after plowing. The soil is loosened and large lumps, or clods, are broken up. Now the rainwater and air can get down into the soil more easily. That is important because all plants need water and oxygen from the air to grow strong and healthy.

The farmer's work with the disk may take several days, too.

At last, he is ready to put the seeds into the earth. Soybean seeds are the soybeans themselves. Every year, some of the soybean crop is saved by the farmer to be used as seed the next spring.

Now the farmer attaches a planter to his tractor. Again he drives slowly across the field. The planter opens grooves in the soil and drops in

Filling a 12-row planter with fertilizer. In front of the fertilizer bins are reserve supplies of soybean seeds. The covered bins are the planters holding the seed. When there are no more seeds in the planters, the farmer can refill from the reserve supplies. In this way, he need not return from the fields as often to get more soybeans.

Spraying before planting to keep weeds from growing. Weeds can choke the soybean plant. On the back of the plow are the curved cutters to break up the soil.

seeds. Then it pushes the earth back over the grooves to cover the seeds.

Planting may also take several days.

As the farmer plants, he sprays the field with weed killer. Or he may have sprayed just before planting, when he disked.

The spring planting is now completed. The farmer hopes the weather will be good. He wants enough sun to keep the earth warm, and enough rain to start the seeds growing. When the seeds start to grow, we say they are *germinating,* or sprouting.

Good weather at that time is very important. A sudden cold snap can weaken the germinating seeds. A sudden heat wave can dry them out as though they were in an oven. Too much rain or too heavy a rain can wash them right out of the ground. Too little rain—called a drought—can make them dry up and die.

But if all goes well, the tiny soybean plants will push up through the earth in about 10 to 14 days after planting. The baby plants, or seedlings, are very delicate. Too much sun can still burn them, and too much rain weaken them. Sudden heat or cold can also kill them. Later on, when the plants grow big and strong, sudden changes in the weather are less likely to bother them.

Now that the seedlings are growing, the farmer

These seedlings have just popped out.

A few weeks later, they look like this.

has another important job to do. It is one that he has to do several times during the early summer, if he has not used weed killers. That job is called *cultivating*.

The farmer attaches a cultivator to his tractor. As he drives through the soybean field, the spikes of the cultivator loosen the earth around the plants and dig up the weeds.

Weeds grow where no human beings planted them. Often they are sturdy, strong plants. They grow fast and spread quickly. If they get too big, they can "choke" the soybean seedlings to death.

If the weather has been good and the farmer has kept the weeds down, his soybean seedlings grow quickly. In six weeks, some kinds of soybeans may be as tall as most first-graders in your school. In ten weeks, they may be as tall as a full-grown person. Other types of soybean plants are small, and never get to be more than three or four feet tall.

Small or tall, all soybeans are very bushy plants. They stand up straight, with stalks that grow strong and as thick as a man's thumb. They put

Cultivating the soil. Some seedlings will be destroyed in this process, but it is necessary to thin out the plants or individual plants will grow too tall and weak, not bushy and strong. ➤

The growing plants. Notice how the leaves are in sets of three.

out branches with many bright green leaves. Soybean leaves always grow in threes.

Most bean plants have leaves in threes. And, of course, soybeans are cousins to the common garden beans like lima beans, kidney beans, stringbeans, and peas.

Soybean leaves, leaf-stems, branches, and later the pods are all covered with short, fuzzy, greenish-gray hairs. These hairs are nature's way of trying to protect the beans against insects. But that protection doesn't always work.

Early in the summer, the soybean flowers begin to appear. They grow in little bunches, or clusters, where the leaf stems come out of the branches. The flowers are usually white, purple, or pink. They have no smell.

The flowers last only about a week. Then they fade, dry up, and drop off the plant. And, here and there, clusters of little, pale green pods will take their place. Day by day, the pods grow longer and hang in bunches. There may be hundreds of

The tiny, lovely, soybean flower.

pods on a single plant. Each pod usually contains two or three soybeans. Sometimes there are more.

As the pods grow, they change color. From pale green, they turn to yellow, brown, dark green, or nearly black. But all the pods on one plant are the same color.

By August, the soybean plants are about as tall as they will get to be. The pods are full-sized, about two or three inches long. They have grown fat and lumpy as the beans inside them have grown.

Now the leaves start to dry. They turn yellow, then brown. They get stiff and papery. Then they fall off. The August sun beats down on the pods. There are no more leaves to act as a sun shade. The beans inside the pods are getting ripe.

Sometime between late August and early October, the soybean farmer is ready to start picking the ripe beans. This is called harvesting.

To harvest the beans, the farmer uses a combine. As the machine is driven through the fields, it does several jobs. It cuts the strong soybean stalks off, close to the ground. Then it picks them up in bunches, and sends them through a whirling cylinder where the beans are separated from

A closeup of the soybean pods and stems with their fuzzy covering, like the skin of a peach.

A farmer checks the ripeness of his plants. Are they ready for harvesting?

their pods. The shelled beans are dropped onto racks which shake them to get rid of any hulls, leaves, or dirt. Finally, the beans are pushed into grain tanks attached to the combine. And the hulls and vines are dropped by the machine onto the ground. All this happens to a soybean plant after just a few seconds inside the combine.

As the grain tanks fill, they are emptied into trucks or wagons. When these are full, the farmer

A combine harvesting a field of soybeans.

The hulled soybeans are poured from the combine into a truck. The farmer will then take the beans to a local storage plant, where they will be sold.

drives to the nearby elevator where he will sell the soybeans. (*Elevator* is the name farmers give to any large building where grain is stored.)

Meanwhile, the harvest has ended. Now the farmer plows the dead stalks into the earth. Along with the soybean roots, the stalks will decay in

From the truck, the soybeans are poured onto the moving steps of a small storage tank.

Again the soybeans are trucked, this time to a buyer. Now the beans will be stored in huge elevators, the ones you see in the background. Moving steps will take the beans to a particular room inside the elevator that is set aside for this species of soybean.

the earth. That will help to keep the soil fertile for the next crop.

And next year, the farmer will wait impatiently for spring to come so he can get the soil ready once more for the soybean seeds.

·5·

Soybeans and Your Health

ONE OF THE MOST HEALTHFUL FOODS you can eat are soybeans. That is because close to half of every soybean is protein.

Proteins are made of many different amino acids. Amino acids are what our bodies use for growth, maintenance, and repair. Eight of these amino acids are essential. We must have them every day for good health.

The foods that give us the most of the eight amino acids are eggs, fish, meat, milk, cheese— and soybeans.

Of all these foods, soybeans are the cheapest. They are also the most quickly grown. They take only 16 to 20 weeks, and so they are more readily

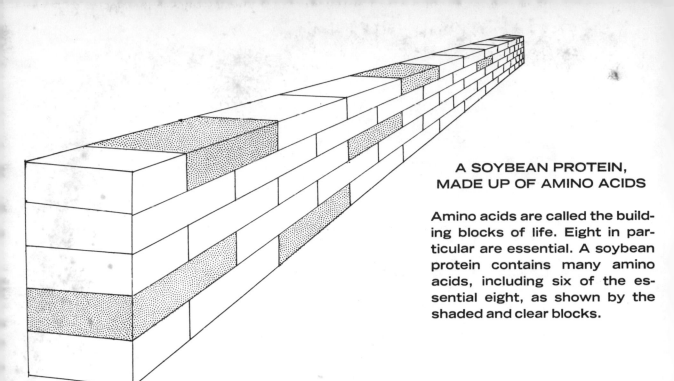

A SOYBEAN PROTEIN, MADE UP OF AMINO ACIDS

Amino acids are called the building blocks of life. Eight in particular are essential. A soybean protein contains many amino acids, including six of the essential eight, as shown by the shaded and clear blocks.

Some of the foods that contain the amino acids your body must have.

52

available. It takes years to raise cattle for beef or milk. It takes many months to raise chickens or fish.

And soybeans can be grown easily in many parts of the world. The farmer doesn't need expensive machines to grow them, though most American soybean farmers do use big machines.

All these things are important right now because food prices are very high. Many people, including some in the United States, are too poor to buy eggs, fish, cheese, milk, and meat. These foods are always the most expensive ones in any country.

What if you don't eat meat, eggs, cheese, or fish? What if you don't drink milk? Can you get the eight special amino acids by eating soybeans every day?

The answer is no. Good as they are for you, soybeans contain only six of the eight special amino acids we need. But the other two are easy to get from rice or some other grain, like corn or wheat. Grains contain the two amino acids that are usually missing in soybeans.

The important thing to remember is that you must eat the soybeans and the grain at the same meal. Your body can get the most out of the eight special amino acids *only* if they are eaten at the same time.

Thanks to the many ways soybeans can be processed, it is easy to eat them every day without getting tired of them. They come in such canned foods as vegetable soup and baked beans. They can be bought dried and, like other dried beans, can be used to make soups, casseroles, and other bean dishes.

Soybeans are also made into soy *analogs*. That is the name given to foods that are processed to look, smell, and taste like other foods, such as

All the meats and sauces in this photograph have been extended with soybeans.

ham, bacon, beef, and chicken. If you have eaten them, were you able to tell the difference between the analogs and the real thing?

Another way that soybeans are eaten is as an *extender*—something that makes meats go farther. Manufacturers of processed foods like hot dogs and meat products use soybean extenders.

All over the world, scientists are working on the use of soybeans as meat substitutes. They are trying to make soy analogs for favorite foods of people in many nations. They are also trying to make soy analogs for foods other than meats.

If they succeed, more people can have healthful and inexpensive foods to enjoy every day of their lives. And soybeans will have helped stamp out malnutrition, hunger, and starvation in the world.

Scientists working on soybean milk production in Indonesia.

·6·

Soybeans in Your Closet

YOU MAY NOT HAVE A PLACE outdoors to plant soybeans. But you can still have the fun of eating soybeans you have raised yourself—right inside your own house or apartment.

To start, buy a bag of dried soybeans in the supermarket. Cover the bottom of a pie plate or cake pan with beans. One layer is all you need.

Cover the beans with warm water—not too hot

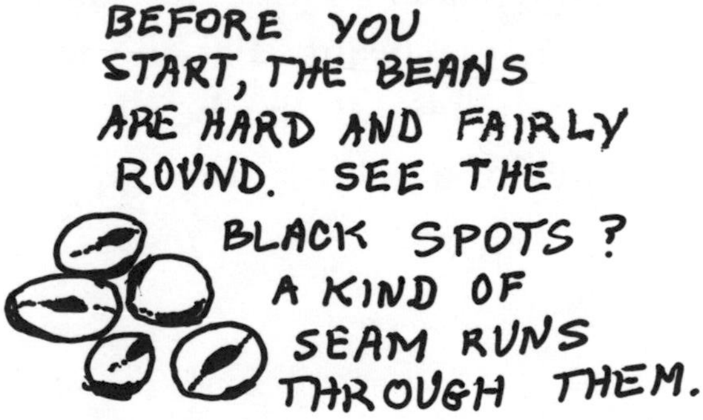

BEFORE YOU
START, THE BEANS
ARE HARD AND FAIRLY
ROUND. SEE THE
BLACK SPOTS?
A KIND OF
SEAM RUNS
THROUGH THEM.

and not too cold. Leave them to soak overnight.

In the morning, the beans will be soft and a little swollen. And instead of being round, they will be sort of boat-shaped.

Pour off all the water. Do not dry the beans.

AFTER SOAKING, THE BEANS LOOK MORE LIKE THIS-- SWOLLEN AND BOAT-SHAPED.

They should still be wet, even though they're not in water. Put the beans back into the plate or pan so there is one layer covering the bottom.

Now take a sheet of paper toweling. Wet it well. Cold water is fine for this. Lay the paper towel over the beans. Pat the towel down gently until it touches *all* the beans. You'll be able to see if they're touching through the top of the wet towel.

THE BEANS ARE IN THE PIE-PLATE, READY FOR THE WET PAPER-TOWEL TREATMENT.

IMPORTANT: If there is a pool of water at the bottom of the plate or pan, pour it off very carefully. Your beans should be wet from the paper towel, but they should not be sitting in water, or they will rot.

Now put the plate in a dark place. If you have a closet under the kitchen sink, that's perfect. If you don't, use another closet that is opened every day. Opening a closet and seeing your beans is a good way to remember that there are still things you need to do!

The most important thing is to check the paper towel on your plate every morning and every night. If it is getting dry, gently sprinkle a little water over it. It's all right to lift the towel to see how your beans are growing. But be sure to pat it down again when you have finished looking. And don't worry about how many times the closet is opened. As long as the beans are *mostly* in the dark, they'll be fine.

It may only take four or five days until something happens, or it may take a few weeks. Every day your beans will seem to be getting softer and a tiny bit more swollen. Then one day, all of a sudden, you'll notice that some of them have split open along a little seam near the black spot.

HERE'S A CLOSE-UP OF A SOYBEAN THAT HAS SPLIT. IT'S GETTING READY TO SPROUT.

AND HERE'S A
TENDER YOUNG SPROUT,
JUST BEGINNING TO
GROW.

Out of the opening, you will see a tiny white tip. That means your beans are germinating. Those tips are sprouts.

Not all the beans will sprout at the same time. Some may not even sprout at all. Just keep the beans in the dark closet under the wet paper towel, and check them twice a day, exactly as you have been doing.

Day by day, your sprouts will grow longer. When they are an inch or two long, they will be ready to eat.

Now throw away the paper towel. Rinse the beans and bean sprouts under running cold water. The old skins that are lying in the plate will wash away. Some skins will be sticking to the sprouts, but water will wash them off. Don't eat the skins; they're tough and tasteless.

But the sprouts will be *delicious!* The best way to eat them is raw. They are crunchy and tender with a sweet flavor.

Many people like to mix the soybean sprouts with a little mayonaise. Add salt and pepper and spread the mixture on whole wheat bread for a sandwich. That's a perfect meal, with all eight of the amino acids you need.

You can keep soybean sprouts in the refrigerator for at least 10 days or two weeks. Just put them in a plastic bag with a few drops of water, seal the bag, and they'll stay fresh and tasty. Or you can put them in a jar filled with water.

For variety, drop a few soybean sprouts in your favorite soup, or mix them with other vegetables —just before you're ready to eat. They're good in salads, too.

Best of all, you can sprout soybeans this easy way all year 'round.

INDEX

A

air, 20, 35
America. *See* United States
amino acids, 51, 53
analogs, 54-55

B

beans, 9, 13, 18, 54. *See also*
 soybeans
butter, 25

C

cakes, 11, 25
Carver, George Washington,
 26
cattle, 19, 25, 53
cereals, dry, 10, 25
cheese, 51, 53
chickens, 53, 55
China, 15, 24, 33
combines, 44-45
cookies, 11, 25
cosmetics, 13, 28
crackers, 25
crop rotation, 21
cultivating, 40

D

detergents, 12
disking, 35, 37

dressings, bottled, 12
drought, 38

E

earth. *See* soil
eggs, 51, 53
elevator, grain, 48
extender, meat, 55

F

fabrics, 13
farmers, American, 15,19, 21
 33-49; Chinese, 15
feed, animal, 13, 19, 21, 24,
 26
fertilizer. *See* green manure
fish, 51, 53
food, 51-55; industry, 28.
 See also named foods

G

germination, 38
grains, 53-54
green manure, 21, 26

H

harvesting, 44, 48

I

ice cream, 11
ink, 12
instant coffee, 12

DISCARDED